Table of Content

Introduction

The idea behind this cookbook is to reproduce a selection of historical recipes. My aim, in writing this book, is to compile a list of recipes from antiquity, the Middle Ages and the Renaissance that you can actually easily do at home, in a normal household kitchen, with readily available ingredients. The broad time frame chosen for this book allows a glimpse of traditions, transitions as well as innovations in historical cooking.

On several occasions I propose substitute ingredients, depending upon availability and taste preferences. There is one important thing we have to understand: a historical recipe is normally just a list of (proposed) ingredients. It usually does not give any amounts, nor does it describe the cooking process in detail. And the ingredients listed are those that were available to the society's upper class, for it is for their tables that recipes were written. Everyone else would make do with what was available to them, often aspiring to imitate the cuisine of the wealthy. And, frankly, I like to re-interpret recipes. All authenticity debates aside, in the end, the purpose is always to cook a tasty meal. Nevertheless, I will avoid proposing anachronistic substitutes; the idea is to always keep in mind what would have been available at the time in the respective parts of the world. We can nowadays hardly imagine cooking entirely without all the things that were unknown in the Old World before their introduction from the Americas: potatoes, tomatoes, chilies, pumpkins, most kinds of beans, corn, chocolate ... They have conquered the kitchens of this planet, and yet rich culinary traditions existed which made do without them: Italy without tomatoes, Germany without potatoes, Asia without chillies? Yes, indeed.

Another consideration is seasonality. Of course, everybody would, until rather recently, eat strictly seasonally. The only exceptions were dried, pickled, cured and fermented foods that were available all year round. Going back to these roots is not only historically correct but also part of a sustainable lifestyle.

Cooking methods have changed with kitchen technology. In Babylonia, earthenware would have been the cooking pot of choice. Pottery remained the common cookware over many centuries, being only gradually replaced by metal pots. If you have a fireplace, you might consider getting a ceramic pot in which stews or pulses can slowly simmer until they're done. A Dutch oven is a good alternative; it is usually made of cast iron, has a shape similar to that of a medieval three-legged cooking vessel, and can be heated entirely with charcoal. Copper pots are an option too. However, my all-time favorite for everyday cooking is the modern stainless steel pot or pan.

What really matters to me is to understand the concept of food in each of the periods described here: What was considered good food? Which tastes and which cooking methods were appreciated? What did a cook have to do to be appreciated as a master chef? And, which eating habits were considered healthy? You will see that throughout history some of the answers to these questions differ vastly from our beliefs. Nevertheless, it is completely possible to recreate historical dishes at your home, to incorporate them into your everyday cooking, or even to organize a lavish theme party.

All recipes serve four people, if not stated otherwise.

Recipes from the Garden of Eden:
The Ancient Near East

The main source for ancient Mesopotamian recipes, apart from the archaeological record, is the collection of three cuneiform tablets from the Old Babylonian period that are now part of the Yale Babylonian Collection, and are hence called the Yale Culinary Tablets. Nothing is known about their provenience but the language and style of writing suggests that they come from Southern Mesopotamia, in what is now Southern Iraq, and date to approximately 1,700 BCE. The very short recipes – surely the scribe had a cook in mind that knew the dish and just needed a reminder of the details – are written in the Old Babylonian dialect of the Akkadian language. Due to damaged passages and unknown words not all of the recipes can be clearly understood. But we get a good overview of the cooking methods, ingredients and spices used (lots of onions, leeks and garlic, in every dish!). However, we have to keep in mind that recipes discovered from before the Renaissance usually describe the dishes of the wealthy, which is why the vast majority of the dishes in this collection are based on meat, and they seem to be – to our modern tastes – quite repetitive. Surprising is the total lack of fish dishes which, between the Euphrates and Tigris, must have played an important role in people's diets. This is confirmed by depictions of fishing scenes on cuneiform seals and other sources. I have tried to select a small number of different dishes that represent the variety of foods that people ate in Babylonia three to four thousand years ago.

Pan-roasted barley flatbread

Barley was one of the most important cereals of ancient Mesopotamia, a source not only for bread, but also for porridges, soups and beer – an important part of people's daily diet. Bread was baked from barley and early types of wheat in the form of unleavened flatbreads in simple clay ovens or on hot stones.

You need: 1 cup barley flour
                1 cup wheat flour
                a pinch of salt
                a little oil for the pan
                sesame seeds (optional)

Combine the flour with about ½ cup of water and salt. Knead until you have a smooth and homogenous dough. If it remains sticky, add a little flour. Form four dough balls and flatten each of them with a rolling pin on a flour-dusted surface. Pan-fry the dough disks on both sides in a pan with the oil until brown. And if you like, sprinkle a little sesame seeds over them.

For extra authenticity: Use an old wheat variety such as emmer, spelt or Khorasan wheat.

Babylonian lamb soup

This lamb soup or stew comes in different but always similar variations. They all start with the preparation of a meat broth from water, fat and meat. It is impossible to say whether it has a more soupy consistency or is firmer, like a stew, as no quantities are given. The semolina thickens the soup/stew.

You need: 2 pounds mutton or lamb from leg or shoulder, in cubes
                oil for frying
                salt
                2 onions
                2 tablespoons semolina
                1 leek
                1 cup full fat yogurt
                2-3 cloves garlic

In a deep saucepan, brown the meat cubes in some oil, together with the chopped onions, adding salt to taste and finally three cups of water. Slowly add the semolina to thicken the soup. Leave to simmer for 45 minutes, then add the leek and continue boiling for another 10 minutes. Meanwhile, peel the garlic, press and whisk it with the yogurt. Blend this mixture into the soup at the end of the cooking time and serve with flatbread.

For extra authenticity: Use animal fat, preferably tail fat from sheep, instead of frying oil.

Elamite Yogurt Soup

The name of this dish, originating from Elam in Southwestern Iran, is "Zukanda". The soup is thickened with blood but we can happily omit this ingredient.

You need:
- 3 cups good meat broth from beef or mutton
- 1 tablespoon dill (fresh or dried)
- chives
- coriander greens (cilantro)
- 1 small leek or 2 spring onions
- 5 cloves garlic
- 2 tablespoons flour
- 2 cups full fat yogurt
- ½ stick (¼ cup) butter

In a saucepan, bring the meat broth to boil. Add a tablespoon of dill, the chopped chives, cilantro, leeks or spring onions, as well as two chopped cloves of garlic. Carefully blend the flour and yogurt until you obtain a smooth paste. Gently stir the mix into the soup, bringing to a boil and leaving it to simmer for 15 minutes.

Meanwhile, chop another three cloves of garlic into fine slices and fry them in butter just until they begin to turn brown. Serve the soup drizzled with the garlic butter and flatbread.

For extra authenticity: Add a few teaspoons of blood to the yogurt-flour mix before stirring it into the soup.

Roots with herbs

I have taken the liberty to interpret the recipe's instruction to add coriander, breadcrumbs, crushed leeks and garlic at the end as a fresh herb sprinkle similar to the Italian *gremolata*, thus simply ignoring the fact that the list mentions blood, too.

You need: 1 pound root vegetables (carrots and/or turnips, beets, parsnips)
1 tablespoon butter
1 onion
1 bunch of rocket (arugula)
½ teaspoon coriander seeds
1 tablespoon semolina
2 spring onions
1-2 garlic cloves
fresh cilantro
a handful of breadcrumbs

Clean and dice the root vegetables and the onion. Bring one cup of water to a boil, then add the butter as well as the roots, the onion, the chopped rocket (keep some for the garnish), crushed coriander seeds, and whisk in one tablespoon of semolina. Leave to simmer for 10-15 minutes. In the meantime, crush or finely chop 1-2 garlic gloves, some cilantro and the two spring onions. Mix them with the breadcrumbs and sprinkle them on top just before serving.

For extra authenticity: Well, if you really want to add blood here... go ahead!

Barley pilaf

I must admit that I somewhat made up this recipe. We know that barley played an important role in the Mesopotamian diet, but it must have been too ordinary for the writers of the culinary tablets to write down. So, I just guess from all the other recipes that we have, what might have gone into it. It might well serve as a side dish to one of the following recipes, or be eaten plain with some yogurt.

You need: 3 cups of meat or vegetable broth
             1 tablespoon butter
             2 cups pearl barley
             1 onion
             1 bunch of rocket (arugula)
             2 spring onions
             garlic chives or 1 clove of garlic
             fresh cilantro

In a deep pot toast the pearl barley in a tablespoon of butter, stirring slowly so as not to burn the grains. Once they are just starting to brown a little, add the chopped onion and continue to stir for another five minutes. Then add the broth, turn down the heat and leave to simmer for about 30 minutes.

Chop the rocket, spring onions, garlic and cilantro and add them just before serving.

For extra authenticity: Have lukewarm, stale, unfiltered beer to wash it down.

Babylonian fish

Not a single fish recipe is listed in the Yale Culinary tablets, although fish surely played an important role in Mesopotamians' diet. I've combined whatever evidence there is about a possible fish dish, and came up with the following. Specimens of the carp family are very common in the Euphrates and Tigris Rivers today, and are therefore commonly eaten in modern-day Iraq. I would nevertheless suggest using whatever fresh fish is available in your part of the world.

The vinegar should be mild and of good quality. Babylonians would have used vinegar made from dates, figs, or beer. If none of these is available, a good apple vinegar should do.

You need: 1 2-pound fish or 1,5 pound fish fillet
¼ cup vinegar
salt
1 tablespoon dried mint
2 cloves garlic
butter for frying

Clean and steam the fish. Or, in the absence of a steaming pot, gently sauté the fish with a little liquid in a pan for 20 minutes (depending on its size). In the meantime fry the chopped garlic and the dried mint in butter until they just start to become crispy.

Once the fish is cooked, turn off the heat. Cut open and, if necessary, remove the bones and arrange it in pieces on a plate. Sprinkle with the vinegar, a little salt and the garlic-mint butter.

For extra authenticity: Use river fish and season with date, fig, or beer vinegar.

Two interpretations of a dish:
Babylonian pigeon pie

The recipe described in the Yale Culinary Tablets is something between a pie and a burger. Therefore I am offering you two different interpretations for the modern kitchen.

You need: 2 cups flour
1 cup full fat yogurt
½ stick butter plus butter for greasing
½ cube baker's yeast
2 pigeons, preferably with heart, liver and gizzards
vinegar, salt
½ leek
2 cloves garlic and 1 onion
2 carrots or turnips ("roots")

First prepare the dough: Combine the flour, two thirds of the yogurt, the butter, yeast and a pinch of salt. Knead the dough (add a little water if necessary) and cover it with a cloth, leaving it to rise for one hour. Preheat the oven to 350° F. In the meantime, clean the pigeons. Put them to boil in a small saucepan with just enough water to cover them, adding salt, a dash of vinegar, one onion, the leek, garlic and the carrots or turnips. Boil until they are well done. When the dough has risen well, knead it once more. Separate it into two parts, rolling each of them out with a rolling pin and placing the first one on the bottom of a round, buttered pie dish. The second part will be used as a lid. Remove the cooked vegetables from the pot and mash them together with the rest of the yogurt. Take the pigeons from the saucepan, remove all bones and separate the meat. Chop the gizzards, heart and liver. Place the vegetable paste, the pigeon meat and the chopped offal into the pie dish and cover the pie with the dough lid. Brush the lid with some soft butter and bake for 25 minutes.

Two interpretations of a dish: Babylonian pigeon burger

The original bread crust would have been created by using heated clay pots layered with dough that were turned upside down. That way the two crusts of the pie were shaped and baked separately, being only assembled just before serving, hence the idea of a burger, but with pan-roasted bread.

You need:
- 3 cups flour
- salt
- 1 cup lukewarm milk
- 1 cube baker's yeast
- 2 pigeons, preferably with heart, liver and gizzards
- vinegar
- ½ leek
- 2 cloves garlic
- 1 onion
- 2 carrots or turnips ("roots")
- 2 dates (optional)

Knead a bread dough from the flour, the milk, yeast and a pinch of salt. Cover it with a cloth and leave it to rise for one hour in a warm place. When the dough has risen well, divide it into eight dough balls and roll them with a rolling pin into small flatbreads. Leave them to rise, covered by a cloth, for another 30 minutes.

In the meantime prepare the pigeons and the vegetables as in the previous recipe. While they are boiling, toast the flatbreads in a pan on both sides until golden brown. Serve the vegetables together with the chopped dates and the pigeon between two pieces of bread (like a burger).

For extra authenticity: Add fish sauce to the dough (yes, really)!

Beer stew with meat

This is a tasty dark stew. Use brown beer, not Pilsen or Export.

You need: 2 pounds mutton or lamb from leg or shoulder, in cubes
frying oil
    2 beetroots with some of the greens
    1 onion
    1 bottle (12 oz) dark beer
    salt
    coriander seeds
    ½ teaspoon cumin
    2 tablespoons semolina
    ½ bunch of rocket (arugula)
    1 small leek or 2 spring onions
    2-3 cloves of garlic
fresh cilantro
chives

In a deep saucepan, brown the meat cubes in some frying oil together with the peeled and cubed beetroots, its greens, and the chopped onion. Deglaze with beer, add salt to taste, some coriander seeds and the cumin. Leave to simmer for one hour. Meanwhile chop the leek or spring onions, half a bunch of rocket and the garlic, add to the stew and continue to cook it for another 15 minutes. Make sure there is enough liquid left in the stew, otherwise add a little water. When cooked, bind the sauce with breadcrumbs and sprinkle the finished stew with chopped cilantro and chives and serve with flatbread.

Mersu – a confection of pistachios and dates

This simple but tasty (and even healthy) confection is a sweet dish made from chopped dried fruits and nuts. (The classic version seems to be dates and pistachios but there is no reason not to experiment with adding dried figs, hazelnuts, or sesame seeds). The recipe has been reconstructed from cuneiform texts that specified the shopping list for the King of Mari (now in Eastern Syria).

You need: 24 dried dates
2 ounces pistachios
1 tablespoon honey or fruit syrup, to moisten
sesame seeds (optional)

Pit and chop the dates very finely. Crush the pistachios in a mortar (or, alternatively, inside a closed plastic bag on a table), and mix both with the honey or fruit syrup. Knead well and shape the paste into little balls. Roll the balls in chopped pistachios or sesame seeds and place the confection in a dish.

Ancient Rome

A vast number of recipes have come down to us, the majority from the cookbook "De re coquinaria" said to be written down, or at least sponsored, by Apicius in 1st century Rome. Another notable source is Cato's handbook to rural life, "De Agri Cultura", from the 2nd century BCE. Both represent the cuisine of wealthy gourmets, hence not necessarily what the average worker in Rome would have eaten. The inhabitants of the *insulae*, Roman apartment buildings with up to six stories in which the majority of the urban population lived, were banned from cooking at home due to the high risk of fire. Most people had their warm meals out, on the street or in taverns. Bread was the fundamental base for everyone's nutrition, pulses were highly valued for their nutritional values (and eaten a lot especially by gladiators and soldiers), while meat remained expensive for the poorer segments of society, and fish even more so, apart from the immediate seaside. The social gap widened, and to cook historical recipes increasingly means that we have to decide who we would like to be. Therefore, I would propose that you and I, dear reader, belong neither to the super-rich, nor to the working classes, but are hard-working but successful merchants, wine merchants, maybe, who import a good Falernian wine from Campania to the city of Rome and the land houses of the rich. This would also enable us to enjoy a good glass of red wine with dinner ourselves, instead of drinking the sour dish water made drinkable only by adding honey and spices, with which most people had to get by.

The most outstanding feature of ancient Roman cuisine is the combination of tastes: sweet, sour, salty and even bitter. To achieve this, the majority of dishes contain most of the following ingredients: wine vinegar or verjuice (the juice of unripe grapes), honey or reduced grape must (*vincotto*), fresh herbs and *garum* or *liquamen*, the indispensable fish sauce.

You can use the Thai Naam Plaa or Vietnamese Nuoc Mam as a substitute or, if you prefer not to add fish sauce to practically half the dishes, use soy sauce. In the recipes I will simply refer to fish sauce. *Garum* is very salty, therefore the recipes using it don't add extra salt. If you are replacing the fish sauce, for example with soy sauce, you might need to add a little salt in the end.

There is frequent mentioning of a spice, laser (no connection to the light rays), which was obtained from an extinct plant called *silphium*, which, again, is not identical to the modern plant species with the same name. It is unclear which family the antique *silphium* belonged to. Asaphoetida was recommended in antiquity as a cheaper alternative, and as this one has an aroma reminiscent of leeks and garlic, I would simply recommend one of them as a handy substitute.

A number of ancient recipes call for rue, a bitter herb appreciated by the Romans. You should use it carefully or simply omit it. Pregnant women should abstain from rue completely.

Conditum paradoxum

Let's start with an aperitif. This spiced wine was believed to be a veritable "opener of the stomach", preparing it for the meal to come, at the same time ascribed with almost medicinal properties.

You need: 1 bottle red wine
                  ¾ cup honey
                  1 ounce crushed pepper
                  1 bay leaf
                  1 pinch of saffron
                  2 dates

In a small saucepan, mix the honey with a cup of the wine. And heat it on a slow fire while, stirring constantly. Add crushed pepper, the bay leaf, the saffron and the dates, and leave to infuse overnight. Strain the mixture through a fine sieve and add the remaining wine.

For extra authenticity: Add a pinch of mastic to the spice mixture and double the amount of pepper. Also, the original text calls for using only the date stones but I honestly don't really get the point of doing so.

Roman bread

The common bread of ancient Rome, baked from wheat and barley, was a round loaf dissected into eight pieces to allow for easy portioning. The yeast used was wine yeast, simply obtained from fresh grape must. The yeast on grapes is the same that allows grape must to ferment into wine. This yeast could be kept throughout the year by preserving a batch of each dough as a starter for the next day. Another way to "catch" wine yeast in summer, if you happen to have a vineyard at hand, is to simply place a bowl with a liquid flour-water mix between the vines, and wait for the wild yeast to infuse the dough, which happens in few hours. In absence of wine yeast you can also simply use baker's yeast.

You need: 4 cups wholemeal flour
1 cube baker's yeast, or wild grape yeast
1 tablespoon concentrated grape must (vincotto)
salt

Bloom the yeast in a little lukewarm water. Combine the flour, yeast, concentrated grape must, pinch of salt and two cups of warm water and knead into a firm dough. Cover the dough and leave it to rise in a warm place until it has doubled in volume.

Preheat the oven to 450° F. Shape two round loafs and place them on a baking tray lined with a sheet of parchment paper, leaving them to rise for another 30 minutes. Cross-cut the bread loafs crosswise, then again diagonally, to obtain eight sections. Bake the bread for about 45 minutes. (You can hear if the bread is done by tapping it underneath – if it sounds hollow it's ready.)

For extra authenticity: Serve with moretum cheese or epityrum olive paste.

Libum – ceremonial cakes

These little breads, or rather cheesecakes, were baked for special occasions as ceremonial offerings. The recipe below is according to Cato's "De Agri Cultura".

You need: 2 cups white wheat flour
             2 cups feta cheese (sheep's cheese)
             1 egg
             8 bay leaves
             honey to drizzle (optional)

Preheat the oven to 400° F. Combine the flour, cheese and egg and knead until you obtain a homogenous dough. Shape the dough into 8 small round cakes and place a bay leaf underneath each one. Bake for 30-40 minutes until golden brown. If you wish, drizzle the ready cakes with some honey.

Moretum cheese paste

There are several versions of this popular herb cheese, three alone from Columella's "De re rustica". The last one, "Gallic moretum", is made with hard cheese and resembles more a pesto than a white cheese spread. According to the anonymous "Appendix Vergiliana", garlic is also added, with which I totally agree.

For Version 1 you need: 1 cup cottage cheese or ricotta
1 handful chopped fresh herbs (chives, thyme, mint, parsley, cilantro,
rocket (arugula), wild mustard, rue...)
ground pepper and salt to taste
1 teaspoon wine vinegar
1 clove garlic
olive oil

Mix all the ingredients and drizzle with olive oil.

For Version 2 you add: crushed walnuts
toasted sesame seeds to sprinkle

For Version 3 you need: 1 cup grated cheese (pecorino or similar)
chopped fresh herbs (chives, oregano, parsley, ...)
4 tablespoons pine nuts
ground pepper
1 teaspoon wine vinegar
olive oil

Work the ingredients from the third version into a pesto-like paste with mortar and pestle (or a blender) and serve with Roman bread.

Cato's olive epityrum

This olive paste makes for a wonderful addition to any antipasto platter. Cato himself recommends it as a condiment with cheese.

You need: 2 cups pitted green or black olives
              2 tablespoons olive oil
              1 tablespoon wine vinegar
              ½ teaspoon coriander seeds
              ½ teaspoon cumin
              1 teaspoon fennel greens
              1 teaspoon mint
              1 teaspoon rue (optional)

Chop the olives and, in a mortar (or blender), grind them into a paste, adding olive oil, vinegar, coriander seeds and cumin. Mix in the chopped herbs, keeping some for decoration. Place into a bowl and drizzle with more olive oil.

For extra authenticity: Pit the olives yourself, because Cato explicitly writes "cut them yourself".

In ovis apalis – Eggs in almond sauce

Soft-boiled eggs are served with this unusual dressing – a perfect starter for a buffet party.

You need: 4 eggs
½ teaspoon ground pepper
lovage (fresh or dried)
½ cup almonds
1 tablespoon honey
1 tablespoon wine vinegar
fish sauce (or soy sauce, if you prefer)

Blanch and soak the almonds for one hour in warm water. In a saucepan, bring water to a boil and cook the eggs for 5 minutes, then run cold water over them.

In a mortar, crush almonds, pepper, lovage, honey, vinegar and fish sauce to a smooth paste. Peel the eggs, cut them in halves, arrange them on a plate and place a spoonful of sauce on top of each half.

For extra authenticity: Instead of boiling, poach the eggs one by one in water with a splash of vinegar.

Gourds in the style of Alexandria

This dish's Latin title – Cucurbitas more Alexandrinum – is often translated "Pumpkin Alexandrian style". But pumpkins and squashes originated in the Americas and were unknown in the old world. Cucumbers or gourds from the genus lagenaria must have been the vegetables they used. I tried this recipe with Sicilian snake gourds or the Apulian barattiere, which are firmer than salad cucumbers but have a cucumber-ish taste. Unfortunately both are not very common outside Southern Italy. In the absence of either of these or a similar gourd species, I propose that you nevertheless substitute them with zucchini.

You need: 2 pounds gourd or zucchini
salt and ground pepper
½ teaspoon cumin
½ teaspoon coriander seeds
1 handful fresh mint
10 dates
2 tablespoons pine nuts
1 teaspoon honey
2 tablespoons wine vinegar
fish sauce (or soy sauce, if you prefer)
olive oil

Slice or cut the gourds or zucchini, depending on their size, into bite-sized morsels and blanch them for 1-2 minutes in boiling water. Drain well and mix with a little salt to taste, cumin, coriander and mint. Chop the dates and mix them with pine nuts, honey, vinegar and oil, and pour the resulting sauce over the vegetables. Sprinkle with pepper before serving.

For extra authenticity: Use date wine instead of chopped dates.

Tisana – gladiators' soup

This dish gives us a good idea what a commoner's everyday potage might have tasted like. Especially the gladiators, who were exercising hard but had slave status, and ate so many pulses that they were nicknamed "lentil-eaters". Instead of dried chickpeas, you can also use canned and instead of dried peas, fresh or frozen. This will reduce cooking time dramatically and you won't need to soak them overnight.

You need: 1 cup each of dried chickpeas and pearl barley
½ cup each of brown lentils and dried peas
½ teaspoon salt
olive oil
1 leek
1 bunch cilantro
1 teaspoon dill
¼ cup fennel greens (or chopped fennel)
2 carrots, diced
1 pound kale (or another type of cabbage, but kale is closest to the antique varieties)
½ teaspoon each of fennel seeds, oregano and lovage
fish sauce (or soy sauce, if you prefer)
chives

Soak the dried chickpeas and peas overnight. The next day, change the water and bring to a boil. Leave to cook for one hour, then add the lentils and the pearl barley, salt, a generous dash of olive oil as well as chopped leeks, cilantro, dill, fennel greens, carrots and kale. Boil until the ingredients are soft. Crush fennel seeds, oregano, lovage. Add fish sauce to the mix, and pour over the soup. Before serving, sprinkle with some additional chopped herbs such as cilantro and chives.

For extra authenticity: add a little asaphoetida and some malva leafs.

Lentils with chestnuts

Chestnuts were used to form a significant part of many people's winter nutrition, especially in mountainous regions, right up until the 19th century, when cheap wheat was made available everywhere. A chestnut pest at the same time destroyed the crops, and nowadays chestnuts are considered a delicacy. You can use the fresh, vacuum-packed, frozen or canned versions. In this dish they add a wonderful sweetness to the lentils.

You need: 1 cup brown lentils
1 cup chestnuts, peeled
½ teaspoon ground pepper
½ teaspoon cumin
½ teaspoon coriander seeds
½ teaspoon dried mint
1 teaspoon chopped rue (optional)
1 spring onion, chopped
2 tablespoons wine vinegar
2 tablespoons fish sauce (or soy sauce, if you prefer)
1 tablespoons honey
olive oil
salt

Boil the lentils in triple amount of salted water until soft, about 30 minutes and salt to taste. Boil the chestnuts (if they are not pre-cooked) until soft and crush them together with the spices, herbs, spring onion, vinegar, fish sauce, honey and olive oil. Pour the mix over the lentils and serve.

For extra authenticity: Use asaphoetida instead of spring onion.

Cauliflower crumble

Cabbage in antiquity looked different from the varieties we are used to – not a firm round head, but rather leaves on a stalk. Therefore recipes call for a separate treatment of the two parts. This is an excellent idea even for contemporary cabbage varieties like cauliflower. I have taken some liberties in the preparation of this dish to create two contrasting textures and thus something less "mashy" than the original (see below).

You need: 1 big cauliflower, or 2 small ones
½ teaspoon coriander seeds
1 shallot, chopped
½ teaspoon cumin
salt and ground pepper
olive oil
a handful of breadcrumbs
1 tablespoon concentrated grape must (vincotto)
fresh cilantro

Divide the cauliflower into stalks and florets. Divide the florets into fine pieces and cut the stalks into finger-sized sticks. In a pan with a little oil, roast the cauliflower florets, together with the coriander seeds, chopped shallot, cumin, salt and pepper, stirring constantly, until they brown and start to get crispy. You can add some extra crunchiness by adding a handful of breadcrumbs. Meanwhile, boil the stalks for 5-10 minutes in salted water. Remove from the heat, place them on a plate and cover them with the floret crumble. Drizzle with concentrated grape must and garnish with cilantro.

For extra authenticity: Cook the florets in wine and mash them with the spices instead of roasting them.

Parsnip purée

Apicius recommends this vegetable purée specifically as a side dish for pork. Why not try it with fake boar?

You need:
2 pounds parsnips
2 cups vegetable stock
olive oil
½ teaspoon cumin
1 teaspoon chopped rue (optional)
concentrated grape must (vincotto)
fresh cilantro
2 spring onions

Peel and dice the parsnips. Boil them in the vegetable stock until soft and mash them together with olive oil, cumin and rue. Garnish with a drizzle of grape must, some fresh cilantro leaves and sliced spring onions.

Sardines in parchment

The fleabane in this recipe can be substituted by rosemary.

You need: 2 pounds fresh sardines
1 teaspoon fresh rosemary, or ½ teaspoon dried
some fresh mint leaves
½ cup pine nuts or almonds, blanched
½ teaspoon ground pepper
1 tablespoon honey
olive oil
concentrated grape must (vincotto)
oregano

Preheat the oven to 350° F. Clean, gut and debone the sardines (that is easier than it sounds: pull the bones up from the tail – they will come out in one piece). Crush rosemary, mint, pine nuts or almonds, pepper and honey. Stuff the fillets with the mixture and wrap the fish in parchment, placing them in an oven-proof dish. Bake the fish for 15 minutes. Open the parchment and douse the fish with olive oil, a little concentrated grape must and sprinkle with oregano.

Fish poached in wine

The fish proposed by Apicius for this dish, "ganonas", is unidentified but the text continues with "...and whatever fish you like", so we'll go for that one. You could use one medium-sized fish per person, like sea bream, or several small ones, like red mullets for example, whole or the filets.

You need: enough fish for 4 people ("whatever you like")
        olive oil
        1 leek, sliced
        2 tablespoons fish sauce
        1 cup white wine
        coriander greens
        ½ teaspoon oregano
        ground pepper and salt to taste
        4 fresh egg yolks

Clean the fish and place it in a big pan together with a good dash of olive oil and the sliced leek. On a medium flame, once the fish and the vegetables start frying, add fish sauce and wine, lower the heat, cover the pan, and leave the fish to simmer for 10 minutes (less if they are very small fish). Then turn them around and leave them to cook for another 10 minutes on the other side. The fish should simmer, not fry.

In a separate bowl carefully mix the egg yolks with some of the hot but not boiling liquid from the pan. Add oregano and pepper to taste, and pour the egg sauce over the fish just before serving.

For extra authenticity: Add one tablespoon of reduced grape must, thus adding sweetness to the dish.

Octopus Roman style

This recipe is kept extremely short. It does not mention the use of olive oil, but I find it hard to imagine that it would not have been considered unusual to drizzle some olive oil over the octopus just before serving, so I added it.

Ginger had just been introduced by long-distance trade with South Asia, and therefore it was an exotic luxury. It would most probably have been known only in its dried and powdered form.

You need: 1 octopus (the size varies a lot – adjust to your needs)
garlic chives
ground pepper to taste
1 teaspoon lovage
1 pinch of ginger powder
4 tablespoons fish sauce
olive oil

Clean the octopus, remove the beak, and immerse it in boiling un-salted water. Leave to simmer for 45 minutes. Turn off the heat and leave it to rest for another 15 minutes.

Remove the octopus from the water, drain, and cut it into mouth-sized pieces. Sprinkle with fish sauce, chives, pepper, lovage and ginger powder, and some olive oil.

For extra authenticity: Use asafoetida instead of garlic chives.

Braised fowl

Apicius proposes this preparation for different types of poultry: crane, duck, chicken or pigeon. I agree with all these, except for the crane.

Choose a fowl that has been bred and fed naturally and has been allowed to grow slowly; industrial turbo-chicken don't resemble in any way what a bird would have tasted like in old times.

In this case I would really prefer soy sauce over fish sauce.

You need: 1 chicken (or a similar-sized fowl)
5 dates
½ teaspoon ground pepper
2 tablespoons mustard
1 tablespoon honey
2 tablespoons soy sauce (or fish sauce, if you prefer)
2 tablespoons wine vinegar
2 tablespoons olive oil
1 tablespoon concentrated grape must (vincotto)
fresh cilantro
fresh mint

Preheat the oven to 350° F. Dress the chicken inside an oven-proof dish that contains with ½ cup of water, placing one pitted date each under the wings, the legs and inside. Coat the chicken with a paste made from pepper, mustard, honey, soy sauce, vinegar, olive oil and grape must. Cover the chicken with a lid or a sheet of parchment paper. Braise the fowl for 1 hour, then remove the lid and cook for another 15 minutes. Sprinkle with fresh cilantro and mint and serve, for example with Roman bread and Alexandrian style gourd salad.

Chicken Parthian style

Parthia was a region in north-eastern Iran. The Parthian Empire was Rome's big competitor in the east, and therefore maybe caused a certain fascination (somewhere between awe and exoticism) among Romans. Why this specific dish is considered particularly Parthian is a matter of debate, as the ingredients and the way of preparation are rather Roman and called for in many other recipes. One theory is that a specific chicken breed from Asia was called for.

You need: 4 chicken legs
oil for frying
2 tablespoons fish sauce (or soy sauce, if you prefer)
1 cup wine
½ teaspoons ground pepper
½ teaspoons lovage (fresh or dried)
½ teaspoons caraway
2 garlic cloves

Marinate the chicken legs for a minimum of three hours in a blend of fish sauce, wine, pepper to taste, lovage, caraway and crushed garlic.

Preheat the oven to 350° F. Remove the chicken pieces from the marinade, brown the legs on all sides in a pan with some oil and transfer to an oven-proof dish. Pour the marinade over them and place the dish in the oven. Leave the chicken pieces to broil for 45 minutes. Sprinkle with more pepper and serve.

For extra authenticity: Use asafoetida instead of garlic.

Stuffed dormice (rabbit, really)

This recipe has almost become the trademark dish of ancient Roman cuisine. The dormice, little rodents that are sometimes still eaten in certain parts of the Balkans, yet a protected species in many other European countries, were bred, fattened, and after slaughtering stuffed with minced meat before roasting them in a terracotta dish. Let's recreate this dish with rabbit instead!

You need: 1 rabbit, cleaned
              ½ pounds minced pork
              ½ teaspoon ground pepper
              2 tablespoons pine nuts
              1 shallot, finely chopped
              2 tablespoons fish sauce (or soy sauce, if you prefer)
              olive oil

Preheat the oven to 350° F. Mix the minced meat with the pepper, pine nuts, chopped onion and fish (or soy) sauce. Stuff the rabbit with the meat, close the cavity with toothpicks or twine, and fry it from all sides in olive oil. Deglaze the hot pan with some water and place the rabbit with the liquid in covered oven-proof dish, braising it in the oven for 1 hour 15 minutes..

For extra authenticity: Use asafoetida instead of the shallot.

Minutal in the style of Taranto

A minutal is a type of ragout. This one is in the style of Taranto in Puglia and uses leeks and sausages.

You need: 1 pound fresh sausages
1 leek, white part only
olive oil
½ teaspoon ground pepper
½ teaspoon oregano
lovage (fresh or dried)
2 tablespoons fish sauce (or soy sauce, if you prefer)
½ cup white wine
1-2 tablespoons breadcrumbs

Cut the leek into fine slices and, in a sauce pan, fry them in olive oil. Add the sausages, pepper, oregano and lovage. Stir, then deglaze the pan with wine. Cover and leave to simmer for 20 minutes. When cooked, bind the sauce with breadcrumbs and sprinkle with more pepper before serving.

For extra authenticity: Use red wine instead of white wine. Red wine would have been more available throughout social classes, while white wine was an expensive rarity.

Sweet and sour pork ragout

Another minutal, this time with pieces from pork shoulder.

You need: 2 pounds pork shoulder, diced
2 shallots, diced
olive oil
½ teaspoon ground pepper
½ teaspoon cumin
½ teaspoon dry mint
½ teaspoon dill
1 cup red wine
2 tablespoons fish sauce (or soy sauce, if you prefer)
honey
1 tablespoon vinegar
8 dried prunes, pitted
1-2 tablespoons breadcrumbs

Sauté the diced shallots and pork in an oiled pan until they start to brown. Season with pepper, cumin, mint and dill and deglaze with wine. Add fish sauce, vinegar, honey and prunes and leave to simmer on low heat for one hour. When cooked, bind the sauce with breadcrumbs and sprinkle with more pepper before serving.

Fake boar

Sometimes you want to impress your neighbors in the villa next door, but you just cannot get hold of wild boar. By marinating pork chops according to this recipe you might be able to create a fake boar dish. Make sure to not use too lean meat. It goes well with the parsnip purée.

You need: 2 pounds pork chops
              ½ cup olive oil
              3 tablespoons soy sauce
              1 cup red wine
              2 shallots, chopped
              2 bay leaves
              1 teaspoon each of peppercorns, rosemary, coriander seeds, thyme and
lovage for the   marinade
              1 tablespoon honey
              1-2 tablespoons breadcrumbs
              ground pepper and salt to taste

Leave the pork chops to marinate overnight in the red wine, soy sauce, shallots, bay leaves and spices. In a pan, fry the pork chops in some of the oil taken from the marinade. Once they start to brown, deglaze the meat with the marinade, add honey, cover it with a lid and leave to simmer for half an hour. Once cooked, bind the sauce with breadcrumbs and sprinkle with more pepper before serving.

For extra authenticity: Use fish sauce instead of soy sauce.

Wild boar roast

This is the real thing. You can use a roast from leg or shoulder.

You need: 1 boar roast of about 3 lbs
              1 teaspoon salt
              1 teaspoon cumin
              1 cup red wine
              1 tablespoon honey
              1 tablespoon concentrated grape must (vincotto)
              ground pepper

Rub the roast with salt and cumin and leave it to marinate over night in the fridge.

The next day, preheat the oven to 350° F. Transfer the roast into an oven-proof dish, add red wine, cover the dish with a lid and roast the meat for two hours, basting it regularly (every half hour or so) with some of the liquid in the dish. Once cooked, pour the liquid into a small cooking pot while leaving the roast outside the oven to rest, yet covered to remain warm. Bring the liquid to a boil, adding honey and concentrated grape must. To serve, pour the sauce over the roast and sprinkle with pepper.

For extra authenticity: Add more cumin to the sauce in the end, although I think that the cumin taste would be overwhelming.

Cheesecake in a terracotta bowl

Cato the Elder wrote down this recipe of a honey-cheesecake called "savillum", baked and served in a bowl.

You need: ½ cup flour
     2 cups white cheese
     ½ cup honey plus some for serving
     2 eggs
     oil for greasing
     2 tablespoons poppy seeds

Preheat the oven to 350° F. Mix flour, white cheese, honey and the egg and knead into a soft dough. Spread the dough onto four flat greased terracotta bowls and bake for 30-40 minutes until gold-brown. Before serving, coat with some more honey and sprinkle with poppy seeds.

Pear pudding

A patina is an egg dish baked in a flat dish. It comes in many varieties: plain, savory or sweet, with fruit or nuts. This is the recipe for a sweet pear patina; the unusual spice combination (for us) adds a pleasant tingle to our palates.

You need:
- 3 pears
- ground pepper to taste
- a pinch of cumin
- 4 tablespoons honey
- ½ cup dessert wine
- a little olive oil
- 4 eggs

Preheat the oven to 350° F. Cut the pears, remove the core and dice them. Stew the pears with pepper, cumin, honey, dessert wine and a little olive oil until soft, then mash them. Blend the stewed pears well with the eggs. Pour the mix into flat oven-proof dishes, and bake them for 15-20 minutes. Drizzle with some more honey and pepper, if you like, and serve.

For extra authenticity: Add fish sauce to the pears. I'm not kidding.

Nut pudding

This is a version of patina with mixed nuts.

You need: 1 cup of mixed nuts and kernels (pine nuts, hazelnuts, walnuts, almonds)

    6 tablespoons honey
    1 cup milk
    4 eggs
    olive oil
    ground pepper

Preheat the oven to 350° F. In a pan, roast the nuts. Crush them coarsely in a mortar and mix them with honey. Blend in milk, eggs, a little olive oil and pepper. Pour the mixture into flat oven-proof dishes, and bake for 15-20 minutes. Drizzle with more honey, if you like, and serve.

For extra authenticity: Again, add fish sauce to the pudding. Seriously.

Dulcia domestica – date confect

The combination of salt, pepper and honey will only surprise those who haven't indulged in Roman cuisine yet. Why bore our tastebuds? The combination is delicious, by the way.

You need: 12 dates
              12 walnut kernels
              ground pepper
              2 tablespoons honey
              sea salt

Pit the dates and stuff each one with one walnut and a pinch of pepper. In a pan, carefully heat the honey and let the dates simmer in it for 5 minutes. Turn the dates to make sure that all sides are coated in honey. Place the dates on a plate, allow them to cool and sprinkle them with a little sea salt.

The poor knights of Rome

This dish is so close to French toast or the English "poor knights of Windsor" that I couldn't resist giving them this nick-name. Every bread-eating culture needs a way to use up stale bread, so they come up with similar solutions.

You need: 4 slices of stale white bread
             1 cup milk
             1 egg
             olive oil for frying
             honey

Remove hard crusts from the bread slices. Beat the egg and milk, then soak the bread in the mixture. In a pan, fry the soaked slices in a little olive oil and drizzle with honey.

From Isfahan to Cordoba:
The Medieval Middle East

A number of cookbooks have come down to us from the medieval Middle East, ranging from the 10th to the 17th century, from Persia to Andalusia. Some of them have such poetic names as "Reliefs of the Tables: about the delights of food and different dishes", "The Substance of Life" or even "The Art of Winning a Lover's Heart". The titles reflect the message: The presentation of a dish is as important as its ingredients – all senses are involved. As you can imagine, this is, again, upper class cooking, some of which is straight out of the court of the Caliphs of Baghdad. The courtly cuisine of the Arab world was strongly influenced by Persian culinary traditions, with many dishes still bearing Persian names centuries after they were introduced to the Arab world. The dishes, as fancy as they might have been, are nevertheless very balanced and feature a rich range of fruits, vegetables and fresh herbs. Meats – predominantly lamb, mutton, chicken and other birds, but also veal and beef – were highly appreciated. Pork, of course, was banned.

As in the Roman and the European medieval kitchens, the careful balance of different flavors was valued highly. We see the use of sugar for the first time now, gradually replacing honey, even if it still was quite expensive. Instead of the omnipresent *garum* in ancient Roman cuisine, we find here a favorite seasoning sauce called "murri", the most common version of which is made of fermented barley (although there is a fish variety, too). Despite its name meaning "bitter", it seems to have been a rather salty and umami-flavored, dark-colored liquid, somewhat similar to soy sauce, and thus replaceable by it.

Bazmawurd

This starter from the court of the Caliph of Baghdad was written down in the 9th century cookbook "Kitab al-Tabikh" ("The Book of Recipes") which lists the personal favorites of a number of caliphs, including the famed Harun al-Rashid. *Bazmawurd* is strikingly simple and resembles a contemporary chicken wrap. You could easily take this sandwich to the office, or serve it, cut into pieces, at a party buffet. The amount of fresh herbs, the chopped walnuts, and not least the chopped lemon give it a special touch.

The ideal bread for the wrap would be *lavash* (Persian flatbread), but wheat tortillas or similar flatbreads are fine as well.

You need: 4 flatbreads (like lavash, tortillas or piadina)
1 chicken breast, cooked
1 lemon
½ cup walnut kernels
fresh herbs (tarragon, cilantro, basil, mint, parsley)
olive oil
salt

Slice the chicken breast and place it on the flatbreads. Peel the lemon and chop it into small pieces. Add the lemon pieces, walnuts, roughly plucked herbs, a little olive oil and salt to taste. Roll the bread up and cut it into slices.

For extra authenticity: Instead of lemon use chopped pieces of citron, a less juicy fruit from the same family with a thick rind commonly used for candying.

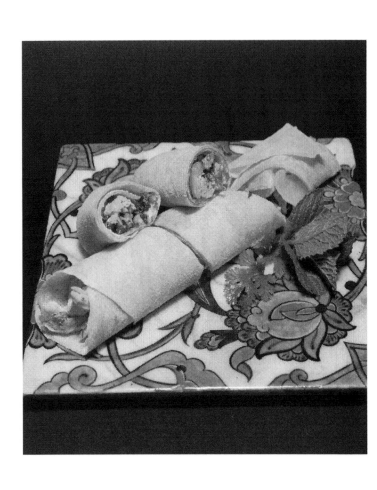

Chicken barida with walnuts and pomegranate

A *barida* (from Arabic *barid* – cold) describes a starter, usually consisting of a cooked base ingredient served with a cold sauce. The recipe is ideal for making use of leftovers.

You need:
1 grilled chicken
2 tablespoons mustard
2 tablespoons wine vinegar
2 tablespoons soy sauce
1 tablespoon sugar
½ cup crushed walnuts
1 tablespoon chopped chives
olive oil
1 pomegranate

Debone the chicken and arrange the pieces on a platter. Blend the mustard, vinegar, soy sauce and sugar into a sauce. Add the crushed walnuts and chives and pour over the chicken. Drizzle the dish with olive oil and sprinkle with the pomegranate seeds.

For extra authenticity: Replace the chives with a pinch of asafoetida and some chopped rue.

Cheese frittata from the oven

This golden frittata can serve as a starter for eight or, with an accompanying salad, a main course for four people.

You need: 2 cups grated or crumbled cheese, preferable goat or sheep
8 eggs plus one extra egg yolk
1 pinch saffron
½ teaspoon ground pepper
½ teaspoon ground coriander seeds
1 pinch clove powder
fresh cilantro
fresh mint
olive oil

Preheat the oven to 400° F. Mix cheese and eggs into a smooth dough, adding all the spices and the chopped fresh herbs (keep some for garnishing). Grease an oven-proof dish with oil and pour the mixture inside. Blend the extra egg yolk with a tablespoon of olive oil and spread it on top to achieve a nice golden surface. Bake the dish for about 20 minutes, until golden-brown, and garnish with some fresh herbs before serving.

Fresh fava beans with hazelnuts

This 13th century recipe transforms fava beans into a filling and delicious salad with tahina, spices, fresh herbs and hazelnuts. Fava beans, also called broad beans, are still very popular in Egypt, albeit it is the dried, soaked and boiled beans that are commonly eaten. I have opted for fresh fava beans here although it is not specified in the recipe, because the recipe that follows in the Kanz cookbook mentions dried beans explicitly, hence no specification might mean fresh beans, cooked only for a short time in salted water. (Fava beans are the only beans that can be eaten raw, which is commonly done here in Italy.)

You need:　2 cups fresh fava beans, shelled
　　　　　　saffron (optional)
　　　　　　½ cup hazelnuts, crushed
　　　　　　1 tablespoon tahini
　　　　　　2 tablespoons vinegar
　　　　　　½ teaspoon each of crushed coriander seeds, rose petals, ginger, and
cardamom
　　　　　　1 pinch each of clove powder and grated nutmeg
　　　　　　fresh mint and parsley
　　　　　　salt and pepper to taste

Boil the fava beans in salted water for five minutes. Drain and color them with a pinch of saffron. Blend all the other ingredients to make a sauce and pour over the beans. Leave the beans to absorb the flavors for some minutes before serving.

Please note that some people suffer from favism, a rare metabolism disorder causing hemolytic response (adverse reaction involving red blood cells) to the consumption of fava beans.

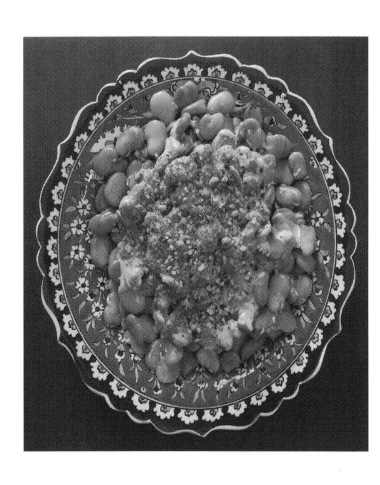

Medieval hummus

We all know the chickpea paste flavored with tahini (sesame paste). This 13th century hummus from Egypt does without it. Instead this recipe aims at a fresh, lemony zestiness, resulting in a paste somewhat lighter than the common contemporary version.

You need: 1 cup dried or 2 cups cooked chickpeas
1 tablespoon wine vinegar
1 lemon pickled in salt, chopped
½ teaspoon cinnamon
½ teaspoon ground pepper
½ teaspoon ginger (powder or freshly grated)
some fresh parsley and mint leaves
olive oil

If you use dried chickpeas, soak them overnight and, the next day, boil them in unsalted water until soft. That might take, depending on the chickpeas, two to three hours. Adding a little bicarbonate to the water speeds up the cooking process.

With mortar and pestle, or in a blender, mix the chickpeas, vinegar, pickled lemon, spices and herbs. Keep a few fresh herbs for the garnish. Arrange the hummus in a bowl, sprinkle with olive oil and the fresh herbs.

For extra authenticity: Add a little chopped, fresh rue as garnish.

Persian barley soup

This soup, an archetype of the "ash-e jo" still popular in Iran, is first mentioned in a 17th century cookbook from the Safavid court.

You need: 2 cups pearl barley
salt
2 cups spinach, chopped
½ cup cilantro, chopped
1 cup unsweetened almond milk
½ teaspoon ground pepper
olive oil

Place the barley in a cooking pot, cover with water and bring to a boil. Add one teaspoon of salt and simmer for about 40 minutes, adding more water if needed. Now add the chopped spinach, cilantro and almond milk and continue to simmer for another 10 minutes. Add pepper to taste and drizzle with a little olive oil.

Itriyya noodle soup

During the time of the Emirate of Sicily and the Arab-Norman era that followed, the island was famous for exporting dried noodles called "itriyya" to its Mediterranean neighbors. In Puglia, "tria", deriving from the word *itriyya*, is still a popular type of pasta, commonly eaten with chickpeas.

You need: 1 chicken
                1 onion, chopped
                2 tablespoons olive oil
                1 cinnamon stick
                1 cup dried chickpeas
                ½ teaspoon each of ground coriander seeds, ground pepper, long pepper,
ground or   grated ginger and (if available) galangal
                1 pinch each of clove powder and grated nutmeg
                10 ounces of fresh or dried tria pasta, or tagliatelle
                4 hardboiled eggs
                grated cheese

Soak the chickpeas overnight. Clean the chicken and cut it into pieces. Put it in a cooking pot, add the olive oil, onion, cinnamon stick, salt and chickpeas. Cover with water and boil until the chickpeas are cooked (that should – depending on the chickpeas – take a couple of hours). Remove the cinnamon and the chicken, debone it and put the meat back into the soup. Season the soup with the spices. Bring back to a boil and cook the pasta in the soup. Meanwhile, arrange one hardboiled and peeled egg per person on a dish, tossing it with grated cheese and pour the noodle soup on top.

For extra authenticity: Add a dash of rosewater to the soup.

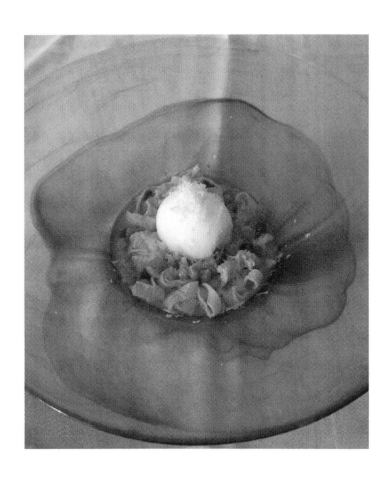

Eggplants with meatballs

Eggplants, originating in Asia, were introduced to the Arab world via Persia. The Arabs then brought them to Southern Europe where they were at first regarded with suspicion, hence the Italian word for eggplant, *melanzana*, from "malasana" – "bad for health". We now fortunately know that eggplants are not unhealthy at all.

You need: 1 pound minced beef
1 shallot, finely chopped
2 pounds eggplants
oil for frying
1 tablespoon tahini (sesame paste)
juice of 1 lemon
1 clove garlic
1 teaspoon sugar
1 handful fresh parsley, cilantro and mint, chopped
salt and ground pepper to taste

Peel and slice the eggplant. Sprinkle it with salt and leave for 20 minutes. In the meantime, add the shallot, some chopped parsley as well as the salt and pepper to the meat and form little meatballs. Fry them in some oil, remove from the pan and set aside. Rinse and dry the eggplant slices, cut them into pieces and, in the same pan, fry them. Once they start to brown, add the meatballs. Mix the tahini with lemon juice and sugar and pour the mix into the pan. Bring back to a boil and season with pressed garlic, fresh herbs as well as salt and pepper to taste.

The eggplants soak much less oil if you dip them in egg white before frying.

Stuffed eggplants

These eggplants are hollowed out and stuffed from the shallow end, not lengthwise, usually done. To do so, choose medium sized eggplants.

You need: 2 pounds eggplants
salt
1 pound minced beef
½ teaspoon coriander seeds
½ teaspoon cumin
½ teaspoon ground pepper
½ teaspoon cinnamon
1 handful fresh parsley and cilantro, chopped
oil for greasing

Preheat the oven to 350° F. Cut the head of the eggplants and set aside. With a small knife and/or a melon baller carefully remove the pulp without cutting the eggplants' skin.

Combine all ingredients, except the oil, and stuff the eggplants with them. Put the eggplants' heads (previously cut off) back and attach them with toothpicks. Carefully place the eggplants into an oven-proof dish greased with oil, sprinkle with a little more oil and bake for about 45 minutes (depending on the size of the eggplants), until done.

For extra authenticity: Pre-cook the meat in sheep tail fat before stuffing the eggplants.

Sardine tartlet

The amounts given are for a starter for four people. To make it a main dish, double the amount.

You need: 1 pound fresh sardines
½ fennel bulb
1 onion
1 handful of fresh cilantro and mint, chopped
olive oil
½ teaspoon cinnamon powder
½ teaspoon ginger, powder or grated
ground pepper to taste

Preheat the oven to 350° F. Clean, gut and debone the sardines (which is easier than it sounds: pull the bones up from the tail – they will come out in one piece). Finely slice onion and fennel. Grease a preferably round oven-proof dish with olive oil, then put one layer of fennel and onion slices together with some herbs, followed by one layer of sardine filets, and so on, until the ingredients have been used up, finishing with a layer of sardines. Drizzle with more olive oil and sprinkle with a little cinnamon, pepper and ginger. Bake for 30 minutes and serve hot.

For extra authenticity: Replace the pepper in the final spice blend with a little mastic.

Fish sikbaj

A *sikbaj* is a dish with a cooked main ingredient, usually fish or meat, served in an acidic vinegar sauce. Our words ceviche, escabeche ("al-sikbaj"), and aspic derive from that term. If available, I would replace the vinegar in this recipe by verjuice (the juice of unripe grapes).

The kind of fish used for this dish *sikbaj* is not further specified; I would recommend white fish like haddock, cod, bass, tilapia or snapper.

You need: 2 pounds white fish fillet
wheat flour
6 tablespoons sesame oil
2 onions, sliced
½ tablespoon each of crushed coriander seeds, ground pepper, cardamom,
and ginger
1 pinch of saffron
¼ cup of mild wine vinegar or verjuice
2 tablespoons honey

Wash and dry the fish, cut it into pieces, and dredge in flour. Fry the pieces in sesame oil. When done, remove them from the pan and set them aside. In the same pan, fry the onion slices until golden-brown. Dissolve the saffron in vinegar (or verjuice) and honey, add the crushed spices and pour the resulting sauce over the fish.

For extra un-authenticity: This fish practically calls for chips. Unfortunately potatoes were not yet known outside the New World. But why not serve fries made from root vegetables like carrots, beetroots or parsnips as a side dish?

Poached fish Andalusian style

This is another fish dish that starts with "take whatever fish you like, big or small, with or without scales". I love that, it really gives you a choice. There is a similar Italian dish called "pesce all'acqua pazza" ("fish in crazy water"), albeit with wine, is usually made from sea bass or bream. This is why I would go for either of them for this dish as well, but other white fish (cod, halibut or sea perch) would do, too.

You need: enough fish for 4 people
             ¼ cup mild wine vinegar or verjuice`
             ¼ cup water
             3 tablespoons soy sauce
             4 tablespoons olive oil
             ½ teaspoon each of ground pepper, coriander seeds, cumin
             1 clove garlic, finely sliced
             1 sprig of fresh thyme, or ¼ teaspoons dried thyme

Scale, clean, gut and rinse the fish. In a pan, mix all the other ingredients and bring to a boil. Add the fish and poach until cooked, which goes fairly quickly (10-15 minutes), depending on the size of the fish. Serve with bread.

Chicken in almond sauce

This sumptuous and unusual preparation might have given contemporary European cooks the idea of a chicken blanc-manger (in the next chapter).

You need: 1 chicken
                sesame oil for frying
                2 cups almond milk
                1 pinch saffron
                2 tablespoons sugar (only if the almond milk is unsweetened)
                ½ cup dried jujubes or apples
                2 tablespoons raisins or sultanas
                rose water
                pistachios
                almonds

Bring salted water to a boil and simmer the chicken in it until almost done (approximately 40 minutes). Meanwhile, in a separate pot, dissolve the saffron and (if necessary) sugar in almond milk. Bring to a boil and leave to simmer until it reduces to about half the amount. Soak dried fruits in the rose water.

Remove the chicken from the water and pat it dry. Divide it into pieces. In a deep pan, fry the chicken pieces in sesame oil on all sides until golden brown. Place the chicken with almond sauce on a dish, arrange the dried fruits on the plate and sprinkle with toasted almonds and pistachios.

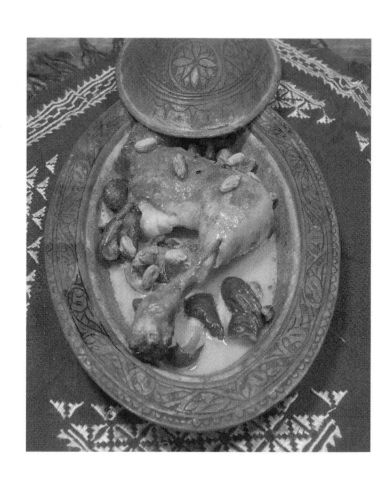

Persian pomegranate chicken on rice

This spectacular dish is said to have been the creation of Safavid Shah Mohammad Khodabandeh, father of Shah Abbas I.

You need: 1 small chicken
                seeds of 1 pomegranate
                1 onion, chopped
                1 tablespoon sultanas
                1 teaspoon ground pepper
                3 cups basmati rice
                salt
                oil or butter for greasing
                ½ teaspoons each of ginger powder, cinnamon, ground pepper
and oregano

Soak the rice in water for at least one hour, then rinse. Preheat the oven to 400° F. Stuff the chicken with pomegranate seeds, chopped onion, sultanas, salt to taste and a good amount of pepper.

Bring the rice to a boil in 6 cups of water with one teaspoon of salt. Once it just starts to boil and the rice is half-done, strain the rice in a colander, and place it inside a well greased oven-proof dish with a lid. Sprinkle with the spice mixture and a little more oil or butter. Place the stuffed chicken on top, cover the dish well with a lid or parchment paper and bake for 1 hour 30 minutes in the middle shelf of the oven. Remove the lid and leave the chicken to brown under the grill. The rice on the bottom should be crunchy.

For extra authenticity: Seal the lid onto the oven dish with some dough.

Lamb sikbaj

Not only fish, but meat as well can be prepared with a sour sauce à la *sikbaj*.

You need: 2 pounds meat from lamb shoulder or leg, cubed
1 cinnamon stick
½ teaspoon coriander seeds
1 teaspoon salt
2 onions, chopped
4 carrots, peeled and sliced
¼ cup of mild wine vinegar or verjuice
2 tablespoons honey
starch for thickening the sauce
4 tablespoons blanched almonds
½ cup dried figs and/or raisins
rose water (optional)

Bring 2 cups of water with salt, cinnamon and coriander seeds to a boil. Add the meat and leave to simmer for 15 minutes. Now add the onions, carrots, vinegar or verjuice and honey. Simmer for another hour while making sure that there is enough liquid in the pot not to burn its contents. Thicken the sauce with a little starch and turn off the heat. Add the almonds and dried fruits and leave the *sikbaj* to cool a little. Serve warm or at room temperature, sprinkled with some rose water, if you like.

For extra authenticity: If available, add jujubes together with the dried fruits.

Tharid – meat stew with bread

This meat stew with bread crumbs soaked in broth is reported to have been the favorite dish of the Prophet Muhammed. Many variations of this popular dish have been created over the centuries. This one is a simple and most likely more original recipe.

You need: 2 pounds mutton meat, cubed
1 cup of dried or 2 cups of cooked chickpeas
2 onions, chopped
1 teaspoon salt
½ teaspoon each of coriander seeds, cumin and ground pepper
1 pound bread (can be stale)
2 tablespoons butter
some fresh cilantro leaves

If using dried chickpeas, soak them overnight and boil them for one hour in fresh water.

Place the meat cubes in a pot, add the chickpeas, onions, spices and salt, and cover with 4 cups of water. Bring to a boil and simmer for a minimum of one hour, until the meat and the chickpeas are soft and tender.

Meanwhile, crumble the bread into small breadcrumbs and place them in a dish. When the meat stew is done, soak the bread in broth and place the meat on top. Pour melted butter and sprinkle with fresh cilantro.

For extra authenticity: Use unleavened flatbread.

Veal couscous

During the Middle Ages, couscous was eaten not only in North Africa but also in Andalusia and Sicily. In Trapani, Sicily, *cuscusu* is still a local speciality and is served with fish, while in North Africa and Andalusia the red meat or chicken recipes (or both) were predominant. You can also prepare this recipe with lamb or chicken.

You need: 1 pound veal, cubed
            oil for frying
            ½ teaspoon each of coriander seeds, cumin, clove powder and ground
pepper
            1 teaspoon salt
            1 onion, chopped
            4 carrots, peeled and sliced
            fresh fava beans, peeled, or green peas
            2 cups instant couscous
            cinnamon powder

In a big pot, fry the meat cubes until brown. Add the spices, onion, carrot, salt and 2 ½ cups of water and bring to a boil. Simmer for 30 minutes, then add the fava beans or green peas and cook for another 10 minutes.

Place the dry couscous in a big dish and pour in 1 ½ cups of the hot veal broth. Wait until the broth has been absorbed, then stir gently and place the meat and vegetables on top. Sprinkle with a little cinnamon powder and serve.

For extra authenticity: You can also steam your couscous in a couscoussier, but this "quick method" was indeed described in 13th century Andalusia.

Sesame seed candy

This sesame honey bar was and remains common throughout the Middle East and the eastern Mediterranean.

You need:　2 cups sesame seeds
½ cup honey
1 tablespoon vegetable oil

Grease a smooth surface, preferably a marble slab, with oil. In a pan, slowly heat the honey and bring to a boil, stirring constantly to make sure it does not burn. Add the sesame seeds, continuing to stir until the mixture thickens. Pour onto the oiled surface and level it. Leave it to cool a little, then, with a knife, make incisions where you want the sesame slabs to be broken later. Leave to cool and harden completely, then break the slabs into pieces.

Crêpes with nuts

What could I say, except "yummy"?

You need: 2 cups wheat flour
2 cups yogurt
3 eggs
salt
vegetable oil for frying
½ cup honey
3 tablespoons crushed walnut kernels
3 tablespoons pistachios
1 tablespoon pine nuts (optional)

Whisk flour, yogurt, eggs and one pinch of salt into a smooth dough. Cover and leave to rest for one hour. Whisk the dough one more time. Heat enough oil in a pan (about one tablespoon per crepe), pour one ladle of dough into the pan, swirl it a little to spread evenly and cook on both sides until just starting to get a golden hue.

In a small pot or pan, heat the honey, add the pistachios and nuts, and pour the mixture over the pancakes.

For extra authenticity: Replace the yogurt with white cheese and a little water.

Bastani – Persian ice cream

This rich ice-cream, a predecessor of which was already eaten in Persia over 2,000 years ago, was refined in the many years that followed.

You need: 6 very fresh egg yolks
½ cup sugar
1 ½ cups heavy cream
1 ½ cups whole milk
1 pinch of salt
1 pinch of saffron
3 tablespoons rosewater
rose petals for garnish
(optional)

Beat the egg yolks with sugar until their color becomes pale. In a saucepan, whisk the cream with milk, salt, and saffron and heat. Slowly pour the hot cream into the beaten eggs, whisking continuously. Pour the mixture back into the saucepan and, on very low heat, stir constantly with a wooden spoon, until the custard thickens. Take care not to boil the mixture. Turn off the heat and continue to stir until the custard has cooled a little. Set it aside and let it cool to room temperature, stirring occasionally. Once it has cooled completely, add the rosewater.

Fill the custard into a metal bowl and put it in the freezer. Take it out and stir it with a fork every half hour, until the ice cream is completely crystallized. Or, if available, use an ice cream machine.

The European Middle Ages

There are quite a number of medieval recipe collections that are known to us, notably from Italy, France, Germany and England. The joy of combining different flavors has remained from antiquity and has even increased. Therefore the distinction between "savory" and "sweet" is sometimes difficult and in some cases even impossible. The medieval table did not know the distinction between savory starters or mains and sweet desserts. Everything was brought to the table at the same time – always assuming that you could afford it, and you could happily nibble on a sweet confection while the chicken just served to you was soaking the slice of bread on which it was placed. You could either eat that bread yourself or leave it as a donation for the poor who, therefore would not only get bread, but also share the flavor of the dishes.

About one third of the year were fasting days, prohibiting the consumption of certain animal products such as meat and, depending on place and time, also eggs and milk products. Fish was always permitted and a vast number of fish and seafood dishes hence became a common treat during Lent. The interpretation of the term "fish" was taken rather loosely: anything that lived in water was a fish in a way, wasn't it? This definition would include beavers, sea mammals and sometimes even water birds.

People ate early: the main meal, dinner, was originally around 10 am and then, in the course of the centuries, slowly moved towards noon, while supper was eaten at sundown. In summer, when the sun set late, a snack was taken in the afternoon. Any additional meal would have been regarded as gluttony. The portions were most probably immense among all levels of society. Work was hard and heating was scarce. The average intake of calories for a medieval farmer is estimated at about 3,000-4,000 calories per day, an intake which was, predominantly made up of cereal and vegetable stews, required huge portions.

Not only mere availability and price determined a food's status and hence who was supposed to eat it, but, also the way it grew was important: the higher the nobler. Hence root vegetables were for the peasants, strawberries would be frowned upon, while tree fruits like apples and pears were a truly noble treat. In general, a lot of attention was given to the appearance of dishes and the whole table. An immaculately clean tablecloth was regarded as indispensable. People ate neatly with their hands, as forks only slowly came into fashion and were at first frowned upon as a way of fussy eating.

The medical theory of "humors" - the four qualities that regulate human bodies: blood, phlegm, yellow bile and black bile – was developed during classical antiquity and was generally regarded as a pillar of medical science throughout the Middle Ages. Foods, too, were categorized that way: hot and dry, hot and moist, cold and dry, and cold and moist. The right balance of these humors was seen as the basis of good health. This concept was the foundation for diets prescribed by doctors to relieve patients from diseases and the imbalances that caused them. The right food humor could also help in daily life. For example, anything hot and spicy like pepper, garlic, onion, ginger or mustard was believed to serve as an aphrodisiac (and was therefore strictly limited in monasteries), while cold and wet foods like cucumbers or pickled vegetables helped to stay abstinent.

Wine-and-cheese-waffles

This nice little recipe is taken from the French household guide book "Le ménagier de Paris" published in 1393. It proposes several different varieties of waffles. I couldn't help but try this seriously tasty cheese and wine version.

You need: 3 eggs
             a pinch of salt
             1 cup red wine
             1 cup flour
             1 cup grated hard cheese
             oil for greasing

Beat the eggs with a pinch of salt, add wine and flour and whisk the dough until smooth. Add the hard cheese. Heat and grease a waffle iron, pour a small ladle of dough onto it, close the iron and bake the waffles until done, which should take about 5-6 minutes.

For extra authenticity: I am using a modern electric waffle maker, but if you happen to have those old-fashioned irons at hand that they would have used in the Middle Ages, you can try to use them. They usually have a beautiful design, too.

Tuscan asparagus

Asparagus was popular around the Mediterranean since antiquity. Due to its shape, it was also regarded as an aphrodisiac. White and purple asparagus, which require a technique of shading each shoot from sunlight with a heap of soil, weren't cultivated yet. Therefore we will use green asparagus.

You need: 3 pounds green asparagus
salt
1 pinch of sugar
4 tablespoons olive oil
2 shallots, finely diced
a pinch of saffron
3 tablespoons white wine
ground pepper to taste
nutmeg

In a deep cooking pot, bring salted water with a pinch of sugar to a boil. Cut off the lower third of the asparagus stems and insert the asparagus, heads up and sticking out, into the boiling water. Boil for 5 minutes, remove and drain. Dilute the saffron in a little white wine. In a pan, heat olive oil and fry the asparagus together with the shallots until they start to brown. Deglaze them with the saffron wine, remove from the heat and serve the dish sprinkled with pepper and a little grated nutmeg.

Onion frittata

Not only tasty but also pretty.

You need: 1 pound red onions
½ cup diced pancetta
olive oil
½ cup grated hard cheese
½ teaspoon ground pepper
salt to taste (depending on the saltiness of the pancetta)
6 eggs
1 pinch of saffron (optional)
a handful of edible flowers to garnish (optional)

Peel the onions and slice them thinly. In a pan, sauté the pancetta in a little olive oil to render its fat. Add the onions and sauté them, stirring carefully.

Beat the eggs with the grated cheese, pepper, salt and saffron. Pour the mixture into the pan, and leave it to cook covered on a low flame until the egg becomes firm. Serve sprinkled with edible flower petals (e.g. borage, nasturtium, calendula, zucchini blossoms, sage flowers...)

For extra authenticity: The anonymous author of this 14th century recipe proposes crocus petals as a garnish. Please note that this means real saffron crocus. There are a number of other flowers also named crocus, some of which are poisonous.

Stuffed eggs

The classic party snack, already in the 15th century.

You need: 8 eggs, hardboiled
½ cup finely grated hard cheese
some leaves of fresh parsley, marjoram and mint, chopped
salt and pepper to taste
a pinch of saffron (optional)
oil for frying
2 tablespoons balsamic vinegar
½ teaspoon ginger powder
a pinch of clove powder
½ teaspoon cinnamon powder
1 tablespoon raisins (optional)

Peel the hard boiled eggs and cut them in half. Remove the yolks and crush half of them, adding the grated cheese, chopped herbs, a little pepper and a pinch of saffron and blend it into a smooth paste. Fill the mixture into the egg whites and fry the egg halves in plenty of oil until golden brown. (A feasible alternative would be to bake them in the oven for 10-15 minutes.) Meanwhile prepare a sauce from the leftover egg yolks. Pass the yolks through a sieve, then blend them with balsamic vinegar, ginger, clove, cinnamon powder as well as some raisins. Serve the fried or baked eggs with the sauce.

For extra authenticity: Use verjuice mixed with a little concentrated grape must instead of balsamic vinegar.

Variations of chicken soup

The anonymous 14th century Liber de Coquina proposes us six variations of a chicken soup. Why not try all of them?

For the base recipe you need: 1 chicken, preferably with its liver
> lard or oil for frying
> 2 onions, chopped
> 1 teaspoon salt

Clean and quarter the chicken. Fry the pieces as well as its liver (if available) in lard or oil together with the onions until brown. Add 4 cups of water and allow to boil for one hour. Remove the chicken bones and divide the meat into small pieces. Pound the liver and put it back into the broth.

For Provence soup add: fresh or dried marjoram, rosemary, parsley, summer savory
> a pinch of saffron
> cinnamon, clove powder, cardamom, grated nutmeg, galangal and ginger
> 1 tablespoon honey
> 4 egg yolks, hardboiled

For chicken soup Provence style, add the herbs and spices as well as the honey and pound the liver together with the egg yolks before putting them back into the soup.

For Martin soup add: ½ cup chopped parsley
> a pinch of saffron
> crumbled bread

For Martin soup, add parsley and saffron and serve on crumbled pieces of bread.

For German soup add: ½ cup chopped fresh parsley, mint, marjoram and rosemary
> a pinch of saffron

For chicken soup German style, add the chopped herbs and saffron.

For French soup add: ½ cup blanched almonds
> 2 cloves garlic
> 4 slices of bread
> ground pepper

For French soup, crush the almonds together with the garlic and blend into the soup. Place a slice of bread into each plate, sprinkle with pepper, and pour the soup on top.

For Saracen soup add: 4 slices of bread
                4 tablespoons verjuice or lemon juice
                ½ teaspoon each of ground pepper and ginger powder
                4 dates, chopped
                1 tablespoon raisins
                ¼ cup blanched almonds

For Saracen chicken soup, cut the bread into small cubes and roast them in oil until crunchy (croutons). Add the other ingredients and serve with the croutons.

For Spanish style add: fresh parsley, rosemary, thyme
                ½ teaspoon ground pepper
                a pinch of grated nutmeg
                4 egg yolks

For Spanish chicken soup, pound the liver together with the herbs and spices, mix them with the beaten raw egg yolks and carefully blend the mixture into the hot soup.

Poor man's sop

A "sop" is a typical poor man's food, a simple meal consisting of a broth made from whatever was available, poured over pieces of stale bread, not wasting anything. The following recipe is just an idea of what might have been a possible combination. Feel free to vary it with whatever ingredients, especially herbs, are available and seem fitting to you.

You need: 2-3 cloves garlic
a few leaves of sage and/or other fresh herbs
1 teaspoon salt
½ teaspoon ground pepper
olive oil
8 tablespoons grated hard cheese
4 eggs
stale bread, cut into pieces

In a saucepan, bring 4 cups of water to a boil, adding the crushed garlic, chopped herbs, salt, pepper and a good dash of olive oil. Simmer for 10 minutes. Meanwhile, boil 4 eggs for 5 minutes; the yolk should still be soft. Rinse in cold water and peel them. Place pieces of stale bread into each bowl, place the grated cheese and one egg each on top and pour the soup over it.

Fava bean purée with caramelized onions

This dish has been around in Italy forever and is still popular in Puglia. The modern version includes potatoes, though, which we will have to omit. While nowadays it is commonly served with fried garlic and herbs from the chicory family, medieval recipes preferred caramelized onions. The latter version is said to have been one of the favorite dishes of Holy Roman Emperor Frederick II (the one who had Castel del Monte in Puglia built).

You need: 4 cups of dried fava beans, shelled
½ cup olive oil
salt to taste
4 red onions
oil for frying
2 leaves of sage, chopped, or ½ teaspoon dried
½ teaspoon pepper
2 tablespoons concentrated grape must or apple butter

Soak the fava beans in water overnight. The next day, rinse the beans and cover them with fresh water (about 4 cups). Bring to a boil. Simmer for 1-2 hours, until soft. Meanwhile, peel and finely slice the onions and slowly fry them, stirring gently. Add sage, pepper and concentrated fruit must and let the onions caramelize on low heat.

Add salt to taste and olive oil to the fava beans and, with a wooden spoon, mash the beans into a soft purée. Serve with the caramelized onions.

Sweet vermicelli

The distinction between sweet and savory dishes, between mains and dessert was not as clear as it seems to many of us today. A sweet dish could easily serve as a main, such as this sugary pasta soup that could be even eaten on fasting days.

You need: 1 pound vermicelli
            2 cups almond milk
            a pinch of saffron
            2 tablespoons sugar

In a big saucepan bring 6 pints of water with one teaspoon of salt to a boil. Heat the almond milk in a separate saucepan. Boil the vermicelli in water until almost done, then drain and finish the cooking process in the almond milk. Dilute the saffron in the almond milk and serve the dish sprinkled with sugar.

Leafy greens in almond milk: a Lent dish

Almond milk was an extremely popular ingredient during lent or one of the many, many other fasting days. Whoever could afford saffron seems to have added it to about everything as well.

You need: 4 pounds leafy greens (spinach, borage, chard, and/or beet leafs)
1 cup almond milk
salt to taste
½ teaspoon pepper
anchovies
a pinch of saffron (optional)

Wash the leaves well, chop them and, in a big saucepan, bring to a boil with the almond milk. Add salt to taste, pepper and, if you like, a little saffron. Leave to simmer for about 10 minutes. Garnish with a few chopped anchovies and serve.

For extra authenticity: Sprinkle the finished dish with a little sugar.

Fish in yellow sauce: another Lent dish

For this dish, "grains of paradise" are proposed as a cheaper alternative to the luxuriant use of pepper. As this spice is hard to get these days, and black pepper has become affordable, we may as well switch back to the use of the latter.

You need:
- ½ cup blanched almonds
- 1 cup white wine
- 4 tablespoons verjuice
- 2 pounds fish fillet of your choice
- oil for frying
- ½ teaspoon each of ground black pepper and ginger powder
- 1 pinch of clove powder
- 1 pinch of saffron
- 1 teaspoonssugar

In a kitchen blender, mix the almonds, white wine and verjuice. Strain the mixture through a cloth to obtain almond wine. Bring it to a boil, add the spices and sugar and leave to simmer until it starts to thicken. Meanwhile cut the fish into slices and fry it in oil. Serve the fish covered in the almond wine sauce.

For extra authenticity: Use "grains of paradise", also called "melegueta" or "alligator pepper", instead of black pepper.

Fish patties

These fish cakes are rather plain. Why not try them with the previous recipe's yellow sauce?

You need: 2 pounds fish of your choice
½ cup finely chopped fresh herbs (parsley, chives, cilantro...)
salt to taste
3 tablespoons flour
vegetable oil for frying

Boil the fish until the bones can be easily removed. Squeeze to remove excess liquid. Pound the fish together with the herbs, some salt and the flour. Shape little patties from the mixture and fry them in oil.

For extra authenticity: The original recipe does not call for any flour. You can omit it but the fish cakes tend to fall apart more easily without it.

Crêpes with spiced wine chicken

This is actually one of my daughter's favorite dishes. The combination of fruit (apples and pears work equally well) and chicken is simply good and the alcohol will simmer away in the cooking process.

You need:
2 cups wheat flour
2 cups milk
3 eggs
1 chicken in pieces or 2 pounds chicken breast
2 tablespoons butter
½ cup white wine
¼ tablespoon each ground pepper, cinnamon, and ginger powder
1 tablespoon honey
2 apples or pears
oil for frying, salt

In a deep bowl whisk first the eggs with milk, then add the flour and a pinch of salt, mixing the ingredients into a smooth dough. Set aside and leave to rest for one hour. Melt one tablespoon butter in a pan and fry the chicken pieces until starting to brown. Add the spices, salt to taste, wine and honey, cover and leave to simmer for half an hour. In the meantime, core the apples or pears and slice them. In another pan, fry the fruit pieces in butter on medium heat, sprinkle with a little cinnamon and set aside. Heat oil in a pan (about 1 tablespoon per crêpe), pour one ladle of dough into the pan, swirl it a little to spread evenly and cook on both sides until starting to get a golden hue. Serve the chicken in the sauce together with the fried apples or pears on top of a crêpe.

For extra authenticity: Place one crêpe into a deep plate, arrange the chicken with sauce and fried fruit on top and cover with a second crêpe.

Blanc-manger

This is the medieval show-off dish of wealth par excellence, the "white dish", made from chicken, almond milk, white sugar and soaked white bread or, even fancier, white rice. Motto: the whiter the better, although there are even multicolored variants with portions of the dish dyed yellow with saffron and egg yolk, or green with herbs. There are so many versions of it that I have tried to extract the classic base recipe of blanc-manger, with several flavoring options given below.

You need: 1 pound chicken breast
             3 cups almond milk
             1 cup white short grain rice
             1 level teaspoon salt
             2 tablespoons butter
             2 tablespoons white sugar
             blanched almonds for garnish

Steam or braise the chicken breast in a little water, taking care it does not change color. Set aside and allow to cool. Boil the rice in almond milk with sugar, salt and butter, adding a little water in the process if necessary. Cut the chicken breast into very fine pieces and blend them well with the rice. The result should be a coarse pudding-like mass. Press it into a mould, turn out of the mould onto a platter, garnish with blanched almonds and serve warm.

Flavoring options to be added to the rice are some powdered ginger, a dash of verjuice or a little rosewater. Or, for the yellow version, add a pinch of ground saffron and a couple of raw egg yolks after cooking.

For extra authenticity: Use capon (castrated rooster) instead of chicken – its meat was valued for being particularly mild and white.

Limonia – lemon chicken

The combination of chicken, lemon and egg yolks is simply gorgeous and also features in the Greek *avgolémono* chicken soup.

You need: 1 chicken, in pieces
4 ounces lardo (fat bacon), cubed
2 onions, chopped
½ teaspoon ground pepper
1 pinch each of grated nutmeg and clove powder
salt to taste
juice of one lemon
3 egg yolks

In a deep pot, render the cubed lardo. Add the chicken pieces and chopped onions and sauté until starting to brown. Add the spices, deglaze with ¾ cup of water and leave to simmer for 30 minutes. Season with salt to taste (depending on the saltiness of the lardo). Turn off the heat and place the chicken pieces onto a plate. Add the juice of one lemon to the broth. In a ladle, whisk the raw egg yolks with some of the broth and incorporate the mixture into the sauce. Serve the chicken covered with lemon sauce.

Variation: This dish is called "raymonia" if pomegranate juice is used instead of lemon juice.

Chicken pie

Pies were a common way to serve meat or vegetable stuffing: they are easy to transport and hence a perfect street food or takeout lunch. The same pie can be made with minced red meat instead of chicken.

You need: 1 chicken, in pieces
½ teaspoon each ground pepper and salt
1 pinch of grated nutmeg
2 tablespoons verjuice or mild vinegar
4 slices of bacon
2 cups wheat flour, plus extra for dusting
5 ounces butter in small pieces
1 egg yolk

Mix the wheat flour, a little salt and the butter with your hands until crumbly. Slowly add ¼ cup cold water, kneading the dough until smooth and homogenous. Cover the dough and leave it to rest in a cool place. Meanwhile cook the chicken until tender. Debone and mix the meat with the spices, salt and verjuice. Set aside. Preheat the oven to 400° F. Divide the dough into two parts and, with a rolling pin, flatten each of them on a flour-dusted surface. Place one of the dough disks into a round oven-proof dish. Place the meat mixture on top and cover first with slices of bacon, then with the other dough disk. Cut off the dough around the rim and roll it out again, cutting out little decorative shapes for decoration. Place them on top of the dough and brush the entire surface with one beaten egg yolk. Bake in the oven for 40 minutes or until golden-brown.

Note: The 15th century cook Martino of Como, also called Maestro Martino, says that the meat from bears was especially adapted to make pies.

Sweet veal

A weird combination, it seems. As this recipe is not accompanied with any vegetables, my 21st century mind suggests serving these meat strips on top of green salad, and suddenly the combination seems a lot less strange.

You need: 1 pound veal
1 tablespoon butter
½ cup white wine
1 tablespoon sugar
1 pinch of clove powder
salt and ground pepper to taste
2 egg yolks

Cut the veal into thin strips. Sauté them in butter, then add wine, sugar and spices and simmer until the veal is soft. Turn off the heat and leave it to cool for 10-15 minutes. Remove the meat and arrange it on top of a plate of green salad. Whisk the broth with the raw egg yolks and pour the sauce over the meat and salad.

Caramelized omelette

Sounds unfamiliar, but isn't it a great idea for breakfast?

You need: 4 eggs
                    1 pinch of salt
                    2 tablespoons powdered sugar
                    juice of ½ lemon

Separate the yolks from the egg white. Whip the egg whites with a pinch of salt until stiff. Carefully incorporate the yolks. In a pan, melt the powdered sugar on medium heat with the lemon juice, stirring gently until starting to brown. Pour the egg mixture on top, cover the pan and leave to cook until firm. Serve with the caramelized side up.

Jerusalem delight

The name alone invoked sentiments of exoticism, paradise and even salvation. Who knew that salvation could be that simple?

You need: 1 cup of dried fruits (figs, peaches, apricots, …)
              2 cups of almond milk

Place the dried fruit and the almond milk into a saucepan and bring to a boil. Simmer on low heat until the almond milk has reduced to about two thirds and the fruits are soft. Serve in small bowls and enjoy paradise.

For extra authenticity: Use peaches only.

Sweet barley porridge

That's more like a peasant's breakfast. Or lunch. Or dinner.

You need: 1 ½ cups pearl barley
2 cups almond milk
1 pinch of salt
2-3 tablespoons honey

Bring almond milk with pearl barley and a pinch of salt to a boil and simmer until the barley is soft. If necessary, add a little water. Sweeten with honey and prepare for a day of tedious field work.

Mistembec - Fritters with honey

Carnival is traditionally the last chance to feast and indulge before the start of Lent. That means that in medieval times people were using up all their stocks of animal fat like lard before it might get rancid. Anything that could be fried was fried, including these lovely little dumplings smothered with honey. Their name, "mistembec" - "messo in bocca" (Italian) "mise en bouche" (French) – means "put in the mouth". Of course you can also use vegetable oil instead of lard.

You need: 1 cup wheat flour
            1 cup wheat starch
            one pinch of salt
            ½ cube baker's yeast
            lard or vegetable oil for frying
            honey for generous drizzling

Dissolve the yeast in one cup of lukewarm water. Whisk it with flour, starch,and a pinch of salt into a smooth dough. Cover and leave to rise in a warm place for about one hour.

Heat the lard or oil and, with two teaspoons, place little balls or strings of dough into the hot fat and leave them to fry until golden-brown. Allow the fritters to drain on a paper towel. Place them into a bowl and drizzle with honey.

Renaissance

Renaissance cooking is not radically different from medieval cuisine. The transition towards a focus on the qualities of each single ingredient, an approach we also appreciate in contemporary cooking, happened slowly. The introduction of ingredients from the New World only slowly affected people's eating habits. Pumpkins, zucchini, beans and maize (corn) were the first to be adopted into Italian cuisine; the latter increasingly to replace other cereals in the preparation of polenta. Cocoa was expensive, tomatoes and potatoes were regarded with suspicion, and chilies were nothing but an exotic oddity. In fact, the tomato, nowadays practically a symbol of Italian cuisine, wasn't commonly used until well into the 18th century. Imported spices became more affordable to a wider public. Cinnamon, in particular, can be found in all kinds of dishes. They certainly liked their cheese, too.

An increasing number of cookbooks are being published, many of them in Italy. I have focused here on the works by Rupert of Nola, a Catalan chef who worked at the Court of Naples in the 16th century, Bartolomeo Scappi, the 16th century cook for several popes, and Martino of Como, also called Maestro Martino. Despite having lived a little earlier, in the 15th century, and thus technically still at the end of the Middle Ages, the book "The Art of Cooking" is considered the first "modern" cookbook and marks the beginning of Renaissance Cuisine. And, for the first time, everyday dishes that would have been eaten by the middle class, were written down, which is why in this chapter you will find a number of recipes for relatively simple soups and hotpots, although they almost all require meat broths. Eating times shifted further towards the end of the day, so that the main meal now could be eaten in the early evening.

Sage fritters

Use fresh sage leaves to make this aromatic snack or starter, and serve hot.

You need: ½ cup white wheat flour
                1 tablespoon olive oil
                2 eggs
                1 pinch of salt
                oil for deep-frying
                about 24 fresh sage leaves

Separate egg yolks from egg whites. Whisk the egg yolks with ¼ cup cold water and olive oil, then add the flour and whisk. Whip the egg whites to a foam with a pinch of salt and carefully blend into the dough mix. Heat the frying oil. Dip the sage leaves into the dough and deep-fry until golden brown.

For extra authenticity: Add a pinch of saffron, some sugar and a little cinnamon to the dough.

Yellow or green Zanzarelli soup

One of those simple and good soups for a cold autumn day.

You need: 4 eggs
½ poundsgrated hard cheese (f.e. parmigiano, grana padano,
pecorino, ...)
1 cup breadcrumbs
3 cups good beef or chicken broth
1 pinch of saffron (optional)
OR finely chopped fresh parsley
ground pepper

Whisk the eggs, grated cheese and breadcrumbs. Bring the broth
to a boil, color it either green with chopped herbs or yellow with
saffron and pour the egg mixture into the pot, stirring gently until
it stiffens. Serve in bowls and sprinkle with ground pepper.

Milky noodle soup

This dish is another example of the combination of different tastes we would not necessarily think of combining in modern cooking. But it is definitely worth a try. The option of either goat's milk or almond milk is given in the original recipe.

You need: 18 ounces of noodles (tagliatelle or similar)
2 cups good chicken broth
1 pinch of sugar
1 cup goat's milk or, alternatively, almond milk
cinnamon to taste
4 tablespoons grated hard cheese (e.g. parmigiano, grana padano,
pecorino,   manchego, ...)

Bring the chicken broth to a boil, add one pinch of sugar and add the noodles. When they are almost cooked, add the goat's milk or almond milk and finish cooking when noodles are done. Serve with grated cheese and sprinkle with cinnamon.

For extra authenticity: Sprinkle with sugar as well.

Melon soup

A refreshing summer soup but, alas, the melon is cooked!

You need: one cantaloupe, not too mature
2 tablespoons butter
2 cups of vegetable stock or chicken broth
2 egg yolks
½ cup of gooseberries or unripe grapes
4 tablespoons grated hard cheese (f.e. parmigiano, grana padano,
pecorino,   manchego, ...)

Cut the melon, remove the peel and seeds and dice its flesh. In a saucepan, sauté it in butter, deglaze with vegetable stock or chicken broth and bring to a boil. Strain the soup through a sieve. Incorporate the beaten egg yolks and add the gooseberries or unripe grapes. Serve with grated cheese.

Cheesy bread soup

Not a balanced meal, I dare say, but if you don't eat this every day...

You need: 12 slices of white bread or toast
             4 cups of good beef broth
             4 cervelat sausages
             1 cup grated hard cheese (e.g. parmigiano, grana padano, pecorino, …)
             sugar, grated pepper and cinnamon to taste
             4 balls of mozzarella cheese

Heat the broth with the cervelat sausages inside. Remove the bread crusts and place one slice of bread each into a bowl. Sprinkle it with grated cheese, a little sugar, grated pepper and a pinch of cinnamon and cover with a slice of mozzarella. Add another slice of bread on top and repeat three times until the ingredients are used up. Pour hot broth over the cheesy bread and serve with the sliced cervelat sausages.

For extra authenticity: Sprinkle with extra sugar and cinnamon on top of the finished dish.

Chickpea hotpot

Hearty and filling, a perfect winter food.

You need:
2 cups of dried chickpeas
½ teaspoon bicarbonate
¼ cup olive oil
1 teaspoon salt
1 tablespoon wheat flour
preferably fresh, otherwise dry rosemary and sage
3-4 garlic cloves
ground pepper to taste

If you use dried chickpeas, soak them overnight and, the next day, boil them in unsalted water until soft. That might take, depending on the chickpeas, two to three hours. Adding a little bicarbonate to the water speeds up the cooking process. Once they are soft, drain and rinse the chickpeas. Place them back into the cooking pot, adding two cups of water, whisking in the flour, olive oil, salt, herbs, chopped garlic and ground pepper and bringing it back to a boil. Simmer for another ten minutes and serve hot.

For extra authenticity: Use red chickpeas, if you can get them.

Eggplant casserole

The eggplants are first boiled, then baked.

You need: 2-4 eggplants (more or less, depending on their size)
2 cups mutton or beef stock
1 onion, sliced
1 cup grated hard cheese (f.e. parmigiano, grana padano, pecorino, manchego, …)
4 egg yolks
1 pinch each of ginger powder and grated nutmeg
a handful fresh parsley

Preheat the oven to 350° F. Peel the eggplants and cut them into slices. In a saucepan, bring the meat stock together with the onion slices to a boil. Place the eggplant slices inside and simmer for ten minutes.

Remove the eggplant and, on top of a bowl or the sink, press them flat between two wooden boards. Blend the egg yolks, grated cheese, spices and herbs, and, in an oven-proof dish, layer the flattened eggplant slices with the mix. Place them into the oven and bake for 15 minutes before serving.

For extra authenticity: You might have guessed it – sprinkle with sugar and cinnamon on top of the finished dish.

Cheese pie

For the dough: 2 cups wheat flour, plus extra for dusting
      ½ teaspoon salt
      5 ounces butter in small pieces
      1 egg yolk
For the filling: 1 pound semi-hard cheese, grated
      2 cups chopped spinach
      1 bunch parsley
      1 teaspoons marjoram (fresh or dried)
      4 eggs
      ½ teaspoon pepper
      a pinch of saffron (optional)
      1 tablespoon soft butter

Mix the wheat flour, a little salt and the butter with your hands until crumbly. Slowly add ¼ cup cold water, kneading the dough until smooth and homogenous. Cover and leave it to rest in a cool place. Meanwhile chop the herbs and mix them with the spinach, grated cheese, eggs, pepper, butter and, if you wish, saffron.

Preheat the oven to 400° F. Divide the dough into two parts and, with a rolling pin, flatten each of them on a flour-dusted surface. Place one of the dough disks into a round oven-proof dish. Place the cheese mixture on top and cover with the other dough disk. Cut the dough around the rim and roll it out again, cutting out little decorative shapes for decoration. Place them on top of the dough and brush the entire surface with a beaten egg yolk. Bake in the oven for 40 minutes until golden brown.

Timballo – pasta gratin

There are many versions of *timballi*; this one is kind of cheese lasagna, but made with long pasta. I like to use hollow pastas, like bucatini or ziti. And, obviously, sprinkled with cinnamon and sugar.

You need: 18 ounces bucatini or ziti pasta
½ stick butter
4 balls of mozzarella cheese
1 cup grated parmigiano cheese
sugar, pepper and cinnamon to taste

Preheat the oven to 400° F. Boil the pasta in water until done and place one quarter of the pasta in coils in a buttered oven-proof dish. Cover with one sliced mozzarella ball, a few flakes of butter, and ¼ of the grated cheese. Sprinkle with a little sugar, pepper and cinnamon. Place another layer of pasta on top and repeat until the ingredients are used up. Bake for 20 minutes in the oven and serve hot.

Cold flounder

Flounder is a flatfish that can be found around the globe. You might also use sole, plaice or a similar flatfish.

You need: 4 small flounders or soles, or 2 pounds of flatfish
             oil for frying
             ground pepper
             salt to taste
             1 lemon
             4 tablespoons mild vinegar

Clean, scale and rinse the fish. Heat oil in a pan and fry it until it starts to get golden brown on both sides, pushing it down to prevent the fish from curling up. Sprinkle with pepper, salt, the juice from one lemon and vinegar and leave to cool down in that mixture. Eat lukewarm or cold with bread and salad.

Tuna casserole

This is a highly unusual way of preparing tuna. Raw I would consider it a tuna tartar but it is then baked in the oven. However, ff the quality of the fish permits, I don't mind having it raw.

You need:
- 2 pounds tuna fillets
- ½ cup blanched almonds
- ¼ cup pine nuts
- olive oil
- juice of 1 orange
- some chopped fresh parsley
- some chopped fresh mint
- ¼ tablespoon each of ground pepper, cinnamon powder, clove powder, and grated nutmeg
- 2 cloves garlic

Preheat the oven to 350° F. Chop and pound the tuna together with the almonds and pine nuts, adding 4 tablespoons of olive oil, the juice of one orange, the chopped herbs and the ground spices. Place the mixture into an oven-proof dish and bake for 20 minutes. Meanwhile, chop the garlic and fry the pieces in some olive oil until just getting crispy, taking care not to burn the garlic. Drizzle the tuna casserole with garlic oil and serve.

For extra authenticity: Use tuna eyes as well – considered a delicacy.

Mussel hot pot

This mussel dish is not so different from other traditional European mussel preparation. All the difference is made by the verjuice, and it is really good.

You need: 8 pounds fresh mussels
¼ cup verjuice
1 teaspoon coarsely ground pepper
½ cup chopped parsley

Clean the mussels of their "beards" and possible impurities. Rinse them well, tapping each mussel that is not closed on a hard surface. If it doesn't close, it means that the mussel is dead and should be discarded. Heat a high saucepan with just enough water to cover the bottom. Place the mussels inside and close the lid. After about 5-10 minutes check the mussels: if they are all open, they are done. (Around 10% of mussels tend to remain closed. Most of them open easily, though, without forcing. Only if they are firmly closed should they be discarded.)

Once the mussels are cooked, douse them with verjuice, ground pepper and chopped parsley. Serve with buttered bread.

Marinated capon

A capon is a castrated rooster. Its meat is considered especially white and tender, although this originates from a time where hens were commonly slaughtered after they stopped laying eggs. Therefore, you can perfectly substitute a capon with a chicken. This recipe requires two cooking steps and marinating overnight.

You need: 1 capon or chicken, or 4 chicken legs
             1 teaspoon salt
             2 cups white wine
             ½ cup vinegar
             ¼ cup reduced grape must (vincotto)
             ½ teaspoon ground pepper
             ¼ teaspoon each of cinnamon, grated nutmeg and coriander
seeds
             1 clove
             1 clove of garlic
             flour for dusting
             lard or oil for frying

Cook the chicken in salted water until it is just half done (about 45 minutes for a whole chicken, 30 minutes if you use only chicken legs). Remove it from the water, cut it into pieces and marinate it overnight in a mixture of wine, vinegar, grape must, spices and crushed garlic.

The next day, remove the pieces from the marinade. Bring the marinade to a boil and reduce it to about half its volume. Flour the chicken pieces and fry them in lard or oil. Serve with the reduced wine sauce.

For extra authenticity: Use lard for frying.

Chicken with green grapes

The idea of half-cooking a chicken before continuing to the final preparation seems to have been common practice. Here is another example, especially lauded for its good digestibility and health benefits for the heart, liver and kidneys. If you can't get hold of green, unripe grapes (for example "ghureh" from Iranian shops) you can also use gooseberries.

You need: 1 chicken in pieces, or 4 chicken legs
             1 teaspoon salt
             oil for frying
             ½ cup unripe grapes or gooseberries
             ¼ cup chopped parsley
             ½ teaspoon ground pepper
             1 pinch of saffron

Boil the chicken pieces in salted water until it is just half done (about 30 minutes). Remove them from the water and pat dry. In a frying pan, heat a little oil and toss the unripe grapes in the oil. Add the chicken pieces and fry until done. Dissolve the saffron in ¼ cup water and deglaze the pan with it. Sprinkle with pepper and parsley.

For extra authenticity: If you somehow manage to come by mint geranium, also called costmary, add some to the dish.

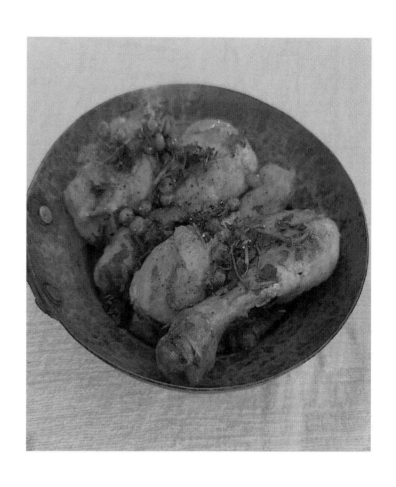

Eggs in sugar

Dessert or not dessert, that is the question. Breakfast, maybe?

You need: 4 eggs
             4 tablespoons sugar
             rosewater (optional)
             cinnamon powder

Place the sugar together with 1 tablespoon of water into a pan and melt the sugar on medium-low heat while stirring gently. Once the sugar has melted and is just starting to caramelize, add the eggs as if frying them in fat. Sprinkle with a little rosewater, if you like, and cover the pan with a lid. Fry the eggs in the caramel for a couple of minutes, then serve sprinkled with cinnamon.

Pinyonada – pine nut cakes

This is a really concentrated dessert; serve small portions.

You need: 1 cup pine nuts
            1 cup blanched almonds
            1 cup sugar

Moisten the pine nuts and almonds with ¼ cup water. Finely pound the almonds and ¾ of the pine nuts in a mortar (or a kitchen blender). Place the pulp in a saucepan, add most of the sugar (leaving a little for dusting at the end) and slowly bring it to a boil, stirring constantly. Once the sugar is well dissolved and starts to caramelize, pour the mixture into small cupcake forms. Sprinkle with the leftover pine nuts and sugar and leave them to cool.

For extra authenticity: Moisten the nuts with chicken stock instead of water.

Zabaglione

Zabaglione (sometimes also spelled zabaione) is still a classic among Italian desserts. Make sure to use very fresh eggs!

You need: 4 egg yolks
             6 tablespoons sugar
             ½ cup Marsala wine or a similar dessert wine

Separate the egg yolks from the whites. In a small saucepan, whip the yolks with the sugar until the cream becomes soft and light. While continuing to whip, add the marsala a little at a time until the cream is well blended. Continue to stir the cream in a bain-marie: Place the saucepan into a bigger pot with hot water and keep on low heat on the stove. Once it gets firm, remove from the hot water bath and continue stirring until the cream cools down a bit. Serve in small glasses.

Bibliography

Some historical recipe collections

Anonymous: Buoch von guoter spise. Würzburg, 14th century.

Anonymous: Cuoco Napoletano. Naples, 15th century.

Anonymous: Kanz. Egypt, 14th century.

Anonymous: Le Ménagier de Paris. Paris, 14th century.

Anonymous (erroneously attributed to Guillaume Tirel, also called Taillevent): Le Viandier. France, around 1320.

Anonymous: Libellus de Arte Coquinaria. Northern Europe, 13th century.

Anonymous: Liber de Coquina. France or Italy, early 14th century.

Anonymous: Libro della Cocina. Tuscany, 14th century.

Anonymous: Manual on Cooking and its Craft. Persia, 16th century.

Anonymous: The Forme of Cury, England, circa 1390.

Anonymous: The Substance of Life. A Treatise on the Art of Cooking. Persia, late 16th century.

Anonymous: The Yale Culinary Tablets. Babylonia, around 1,700 BCE.

Apicius: De Re Coquinaria. Rome, 1st century CE, edited in its current version in the 3rd or 4th century CE.

Cato maior: De Agri Cultura. Rome, 2nd century BCE.

Columella, Lucius Iunius Moderatus: De Re Rustica. Roman Empire, 1st century CE.

Como, Martino of: Liber de Arte Coquinaria. Italy, 15th century.

Ibn Al-Adim: The Art of Winning a Lover's Heart. Syria/Egypt, 13th century.

Different   Ibn Razin al Tujibi: Reliefs of the Tables: About the Delights of Food and the Dishes. Murcia (Andalusia), 13th century.

Ibn Sayyar al-Warraq: The Book of Dishes, Baghdad, 10th century.

Master Chiquart: Du fait de Cuisine. France, 1420.

Naples/   Nola, Rupert of: Libre de doctrina per a ben server, de taller y del art de coch. Barcelona, 16th century.

Nurollah: Madatol Hayat - The Substance of Life. Isfahan, 17th century.

Platina (Bartolomeo de Sacchi): De Honestae Voluptae et Valitudine. Rome/Venice, 1474.

Scappi, Bartolomeo: Opera dell'arte del cucinare. Rome, 1570.

Printed in Great Britain
by Amazon

67677767R00121